M. K. Jones

2017

Algebra Problems

Part 1 (Linear Equations)

Workbook

version 1

(alternative use: keep the workbook blank, and simply photocopy the pages you want to use)

(blank page)

Algebra Problems

Part 1 (Linear Equations)

Workbook

version 1

This workbook belongs to: ______________________________

Date workbook received: ____________

School: ______________________

Grade: ______

Class: ___________

Teacher: _______________

Date workbook completed: _____________

Score: _______

(blank page)

Written Vocabulary Exercise for Mathematics

Student Name: ______________________ Date: __________ Score: ______

addition	
additive inverse	
algorithm	
binomial	
coefficient	
composite	
cube	
digit	
distance	
distributive property	
dividend	
division	
divisor	
equation	
exponent	
factor	
factorial	
factorization	
fraction	
interval	
inverse	
linear equation	
minus	
monomial	
multiplication	
multiplicative inverse	
negative	
number	
numeral	
plus	
positive	
prime	
product	
quadratic equation	
quotient	
reciprocal	
remainder	
square	
subtraction	
sum	
trinomial	
unity	
zero principle	

(end of document)

(blank page)

Algebra Problems 1001
(Algebra Problems: Set # 1001)

Student Name: ______________________________ Date: ____________ Score: _______

Example
Evaluate the following expression, using the 'pull-back' technique.
$527 \times 13 - 527 \times 3$

$527 \times 13 - 527 \times 3 = 527(13 - 3) = 527(10) = 5270$

The 527, occurring in two locations, was 'pulled-back' to one location, with the residues being put into parentheses.

Task 1
Evaluate the following expression, using the 'pull-back' technique.
$534 \times 14 - 534 \times 4$

Task 2
Evaluate the following expression, using the 'pull-back' technique.
$917 \times 12 - 917 \times 2$

Task 3
Evaluate the following expression, using the 'pull-back' technique.
$229 \times 11 - 229 \times 1$

Task 4
Evaluate the following expression, using the 'pull-back' technique.
$723 \times 15 - 723 \times 5$

Task 5
Evaluate the following expression, using the 'pull-back' technique.
$327 \times 6 + 327 \times 4$

Task 6
Evaluate the following expression, using the 'pull-back' technique.
$342 \times 5 + 342 \times 5$

(end of document)

(blank page – can be used as additional scratch area)

Algebra Problems 1002
(Algebra Problems: Set # 1002)

Student Name: ______________________________ Date: ____________ Score: _______

Example
Evaluate the following expression.
$x + 5$, for $x = 3$

If $x = 3$, then $x + 5 = 3 + 5 = 8$

Example
Evaluate the following expression.
$x - 3$, for $x = 20$

If $x = 20$, then $x - 3 = 20 - 3 = 17$

Task 1
Evaluate the following expression.
$x + 7$, for $x = 2$

Task 2
Evaluate the following expression.
$x + 15$, for $x = 5$

Task 3
Evaluate the following expression.
$x + 12$, for $x = 1$

Task 4
Evaluate the following expression.
$x + 1$, for $x = 10$

Task 5
Evaluate the following expression.
$x - 3$, for $x = 10$

(end of document)

(blank page – can be used as additional scratch area)

Notation: Logical Connectives

Definition 1. The symbol "$\wedge$" means "and". (For example, "$x = 2 \wedge x$ is a prime." means "x = 2 and x is a prime.")

Definition 2. The symbol "$\vee$" means "or". (For example, "x is an even number $\vee$ x is a prime." means "x is an even number or x is a prime.")

This sense of "or" allows both conditions to be true. It means that at least one condition is true, and that it is possible that in fact both conditions are true. In the example given above, the number 2 in fact satisfies both conditions. In many cases, however, it will be obvious that only one condition can be true, as in the statement "$x = 2 \vee x = 3$". (x cannot be both 2 and 3 at the same time.)

Definition 3. If each of S and V is a statement, then "$S \rightarrow V$" means "S implies V" (synonymously: "if S, then V"; "S only if V"; "V if S") (For example, $x = 5 \rightarrow x$ is a prime.)

Axiom 1. If S is a statement, then $S \rightarrow S$. (Colloquially, "Every statement implies itself.")

Definition 4. If each of S and V is a statement, then "$S \leftarrow V$" means "$V \rightarrow S$".

Definition 5. If each of S and V is a statement, then "$S \leftrightarrow V$" means ""$S \rightarrow V$" $\wedge$ "$S \leftarrow V$"", and is read "S if, and only if, V". (synonymously: "S is inferentially equivalent to V") (For example, $x = 2 \leftrightarrow x$ is an even prime.)

Definition 6. A tautology is a statement that is always true. (For example, the statement "x = x" is always true, and is therefore a tautology.)

The letter "T" is customarily reserved for a statement known to be true, and the letter "F" is customarily reserved for a statement known to be false.

Definition 7. A set is an entity determined solely by membership. (For example, the set containing precisely the members 5 and 7 is equal to the set containing precisely the members 7 and 5.) (A member of a set is synonymously called an element of the set.)

In some cases, a set may be given in roster form (that is, list form). Roster form is given by explicitly listing the members, in any order, between curly braces, for example, {5, 7}. Note that {5, 7} = {7, 5}. Infinite sets, and finite sets that are very large, can be specified only by description, for example, "the set of primes". The formal way of specifying a set by means of description is given by this example: $\{x \mid x \text{ is a prime}\}$, which is the formal way of specifying the set of primes. It is read, "the set to which x belongs if, and only if, x is a prime", or, more informally, "the set of x such that x is a prime".

Set membership is not affected by redundant entries in a set given in roster form. For example, the set {7, 5, 5, 7, 5} is equal to the set {5, 7}. (When the term "set" is used in a way in which redundant entries are significant, the term "set" is an abbreviation for an "indexed set", sometimes called a "multiset".)

Membership in a set is denoted by the symbol "$\in$" (read "epsilon"). When used in this way, "$\in$" is read "is a member of". For example, "$5 \in H$" is read "5 is a member of H".)

Definition 8. A set containing exactly one element is called a singleton. (For example, the set of even primes is a singleton, because it is equal to the set {2}.)

(blank page)

Notation: Logical Connectives

Definition 9. A set A is said to be a subset of a set B, denoted by "$A \subseteq B$" if, and only if, every member of A is a member of B. (synonymously: $x \in A \rightarrow x \in B$)

"Q.E.D." is an expression from Latin meaning "that which was to be proved".

Theorem 1. Every set is a subset of itself.
Proof: Suppose A is a set. By Axiom 1, $x \in A \rightarrow x \in A$. Thus, by Definition 9, $A \subseteq A$. Q.E.D.

Definition 10. A set A is said to be a proper subset of a set B, denoted by "$A \subset B$" if, and only if, $A \subseteq B$ and $A \neq B$.

Definition 11. A set is said to be infinite if, and only if, it can be put into one-to-one correspondence with a proper subset of itself. (For example, the set of counting numbers 1, 2, 3, ... is infinite, because it can be placed in one-to-one correspondence with the positive even numbers 2, 4, 6, ...)

Definition 12. The set containing no members is called the empty set (synonymously: the null set).

Definition 13. The symbol "$\varnothing$" denotes the empty set.

Since the empty set in roster notation is { }, we have $\varnothing = \{\ \}$.

Theorem 2. The null set has no proper subset.
Proof: Suppose that A is a proper subset of $\varnothing$. Then $\varnothing$ contains a member x that does not belong to A. But then $\varnothing$ is not empty, a contradiction. Thus, $\varnothing$ has no proper subset. Q.E.D.

Definition 14. The primal singleton is $\{\varnothing\}$.

Axiom 2. If F is a false statement and S is any statement, then $F \rightarrow S$ is a true statement. (Colloquially, "A false statement implies anything.") (This axiom is known as the Axiom of Material Implication.)

Theorem 3. The null set is a subset of every set.
Proof: Suppose that A is a set.
Case 1. $A = \varnothing$. Then by Theorem 1, $\varnothing \subseteq A$.
Case 2. $A \neq \varnothing$. Suppose $x \in \varnothing$. Since $\varnothing$ has no members, "$x \in \varnothing$" is a false statement. Then by the Axiom of Material Implication, this implies anything. In particular, this implies "$x \in A$". Since $x \in \varnothing \rightarrow x \in A$, then, by Definition 9, $\varnothing \subseteq A$. Q.E.D.

Axiom 3. If T is a true statement and S is a statement, then $S \rightarrow T$ is a true statement. (Colloquially, "Anything implies a true statement." In particular, a false statement implies any given true statement. This fact has great propaganda value, as the naive realist believes that $(P \rightarrow T) \rightarrow P$. However, even the devil tells the truth when it suits his purpose.)

(end of document)

(blank page)

Algebra Problems 1003

(Algebra Problems: Set # 1003)

Student Name: ________________________________ Date: ____________ Score: _______

Solve for x in the following equations.
(Do not mark up the equation. Use the blank area of this page for your scratch work.)

1. $x + 1 = 5 \leftrightarrow x =$ __________.

2. $x - 1 = 10 \leftrightarrow x =$ __________.

3. $x + 2 = 5 \leftrightarrow x =$ __________.

4. $x - 2 = 20 \leftrightarrow x =$ __________.

(end of document)

(blank page – can be used as additional scratch area)

Algebra Problems 1004
(Algebra Problems: Set # 1004)

Student Name: ______________________________ Date: ____________ Score: _______

Solve for x in the following equations.
(Do not mark up the equation. Use the blank area of this page for your scratch work.)

1. $x - 1 = 14 \leftrightarrow x =$ __________.

2. $x - 1 = 50 \leftrightarrow x =$ __________.

3. $x + 2 = 90 \leftrightarrow x =$ __________.

4. $x - 2 = 45 \leftrightarrow x =$ __________.

(end of document)

(blank page – can be used as additional scratch area)

Algebra Problems 1005
(Algebra Problems: Set # 1005)

Student Name: ______________________________ Date: ____________ Score: _______

Example
Evaluate the following expression.
5x, for x = 4

If x = 4, then 5x = 5(4) = 20

Example
Evaluate the following expression.
x/2, for x = 12

If x = 12, then x/2 = 12/2 = 6.

Task 1
Evaluate the following expression.
5x, for x = 7

Task 2
Evaluate the following expression.
8x + 15, for x = 2

Task 3
Evaluate the following expression.
2x, for x = 4

Task 4
Evaluate the following expression.
3x, for x = 6

Task 5
Evaluate the following expression.
x/2, for x = 10

(end of document)

(blank page – can be used as additional scratch area)

Algebra Problems 1006

(Algebra Problems: Set # 1006)

Student Name: ________________________________ Date: ____________ Score: _______

Solve for x in the following equations.
(Do not mark up the equation. Use the blank area of this page for your scratch work.)

1. $2x = 10 \leftrightarrow x =$ __________.

2. $4x = 44 \leftrightarrow x =$ __________.

3. $2x = 18 \leftrightarrow x =$ __________.

4. $10x = 30 \leftrightarrow x =$ __________.

(end of document)

(blank page – can be used as additional scratch area)

Algebra Problems 1007
(Algebra Problems: Set # 1007)

Student Name: ________________________________ Date: ____________ Score: _______

Solve for x in the following equations.
(Do not mark up the equation. Use the blank area of this page for your scratch work.)

1. $3x = 12 \leftrightarrow x =$ __________.

2. $10x = 40 \leftrightarrow x =$ __________.

3. $3x = 15 \leftrightarrow x =$ __________.

4. $3x = 9 \leftrightarrow x =$ __________.

(end of document)

(blank page – can be used as additional scratch area)

Algebra Problems 1008
(Algebra Problems: Set # 1008)

Student Name: ______________________________ Date: ____________ Score: _______

Example
Evaluate the following expression.
$2x + 1$, for $x = 7$

If $x = 7$, then $2x + 1 = 2(7) + 1 = 14 + 1 = 15$

Example
Evaluate the following expression.
$2x - 3$, for $x = 25$

If $x = 25$, then $2x - 3 = 2(25) - 3 = 50 - 3 = 47$

Task 1
Evaluate the following expression.
$2x + 1$, for $x = 3$

Task 2
Evaluate the following expression.
$2x + 1$, for $x = 12$

Task 3
Evaluate the following expression.
$2x + 1$, for $x = 8$

Task 4
Evaluate the following expression.
$2x + 1$, for $x = 10$

Task 5
Evaluate the following expression.
$2x - 3$, for $x = 12$

(end of document)

(blank page – can be used as additional scratch area)

Algebra Problems 1009

(Algebra Problems: Set # 1009)

Student Name: ______________________________ Date: ____________ Score: _______

Solve for x in the following equations.
(Do not mark up the equation. Use the blank area of this page for your scratch work.)

1. $2x + 1 = 15 \leftrightarrow x =$ __________.

2. $2x - 1 = 19 \leftrightarrow x =$ __________.

3. $2x + 1 = 41 \leftrightarrow x =$ __________.

4. $2x - 5 = 35 \leftrightarrow x =$ __________.

(end of document)

(blank page – can be used as additional scratch area)

Algebra Problems 1010
(Algebra Problems: Set # 1010)

Student Name: ______________________________ Date: ____________ Score: _______

Solve for x in the following equations.
(Do not mark up the equation. Use the blank area of this page for your scratch work.)

1. $2x + 3 = 19 \leftrightarrow x =$ __________.

2. $2x - 3 = 25 \leftrightarrow x =$ __________.

3. $2x + 1 = 11 \leftrightarrow x =$ __________.

4. $2x + 3 = 45 \leftrightarrow x =$ __________.

(end of document)

(blank page – can be used as additional scratch area)

Algebra Problems 1011

(Algebra Problems: Set # 1011)

Student Name: ______________________________ Date: ___________ Score: _______

Solve for x in the following equations.
(Do not mark up the equation. Use the blank area of this page for your scratch work.)

1. $3x + 1 = 22 \leftrightarrow x =$ __________.

2. $3x - 2 = 37 \leftrightarrow x =$ __________.

3. $3x - 2 = 13 \leftrightarrow x =$ __________.

4. $3x + 1 = 16 \leftrightarrow x =$ __________.

(end of document)

(blank page – can be used as additional scratch area)

Algebra Problems 1012
(Algebra Problems: Set # 1012)

Student Name: ______________________________ Date: ____________ Score: _______

Solve for x in the following equations.
(Do not mark up the equation. Use the blank area of this page for your scratch work.)

1. $5x + 1 = 11 \leftrightarrow x =$ __________.

2. $5x - 7 = 43 \leftrightarrow x =$ __________.

3. $2x + 7 = 57 \leftrightarrow x =$ __________.

4. $5x + 7 = 12 \leftrightarrow x =$ __________.

(end of document)

(blank page – can be used as additional scratch area)

Algebra Problems 1013
(Algebra Problems: Set # 1013)

Student Name: ______________________________ Date: ____________ Score: _______

Example
Solve for x: $2x + 5 = x + 15$

copy the equation	$2x + 5 = x + 15$
subtract x from both sides	$x + 5 = 15$
subtract 5 from both sides	$x = 10$

Task 1
Solve for x: $2x + 3 = x + 5$

copy the equation	
subtract x from both sides	
subtract 3 from both sides	

Task 2
Solve for x: $2x + 8 = x + 9$

copy the equation	
subtract x from both sides	
subtract 8 from both sides	

Task 3
Solve for x: $2x + 4 = x + 7$

copy the equation	
subtract x from both sides	
subtract 4 from both sides	

Task 4
Solve for x: $2x + 1 = x + 21$

copy the equation	
subtract x from both sides	
subtract 1 from both sides	

(end of document)

(blank page – can be used as additional scratch area)

Algebra Problems 1014
(Algebra Problems: Set # 1014)

Student Name: ______________________________ Date: ____________ Score: _______

Example
Solve for x: $5x + 8 = 3x + 28$

copy the equation	$5x + 8 = 3x + 28$
subtract 3x from both sides	$2x + 8 = 28$
subtract 8 from both sides	$2x = 20$
divide both sides by 2	$x = 10$

Task 1
Solve for x: $5x + 6 = 3x + 14$

copy the equation	
subtract 3x from both sides	
subtract 6 from both sides	
divide both sides by 2	

Task 2
Solve for x: $5x + 10 = 3x + 20$

copy the equation	
subtract 3x from both sides	
subtract 10 from both sides	
divide both sides by 2	

Task 3
Solve for x: $5x + 5 = 3x + 7$

copy the equation	
subtract 3x from both sides	
subtract 5 from both sides	
divide both sides by 2	

(end of document)

(blank page – can be used as additional scratch area)

Algebra Problems 1015
(Algebra Problems: Set # 1015)

Student Name: ______________________________ Date: ____________ Score: _______

Example
Solve for x: $x + 3 = 23 - x$

copy the equation	$x + 3 = 23 - x$
add x to both sides	$2x + 3 = 23$
subtract 3 from both sides	$2x = 20$
divide both sides by 2	$x = 10$

Task 1
Solve for x: $x + 8 = 18 - x$

copy the equation	
add x to both sides	
subtract 8 from both sides	
divide both sides by 2	

Task 2
Solve for x: $x + 5 = 13 - x$

copy the equation	
add x to both sides	
subtract 5 from both sides	
divide both sides by 2	

Task 3
Solve for x: $x + 1 = 15 - x$

copy the equation	
add x to both sides	
subtract 1 from both sides	
divide both sides by 2	

(end of document)

(blank page – can be used as additional scratch area)

Algebra Problems 1016
(Algebra Problems: Set # 1016)

Student Name: ______________________________ Date: ____________ Score: _______

Example
Solve for x: $x - 3 = 27 - 2x$

copy the equation	$x - 3 = 27 - 2x$
add 2x to both sides	$3x - 3 = 27$
add 3 to both sides	$3x = 30$
divide both sides by 3	$x = 10$

Task 1
Solve for x: $x - 1 = 11 - 2x$

copy the equation	
add 2x to both sides	
add 1 to both sides	
divide both sides by 3	

Task 2
Solve for x: $x - 1 = 2 - 2x$

copy the equation	
add 2x to both sides	
add 1 to both sides	
divide both sides by 3	

Task 3
Solve for x: $x - 8 = 22 - 2x$

copy the equation	
add 2x to both sides	
add 8 to both sides	
divide both sides by 3	

(end of document)

(blank page – can be used as additional scratch area)

Algebra Problems 1017

(Algebra Problems: Set # 1017)

Student Name: ________________________________ Date: ____________ Score: _______

Solve for x in the following equations.
(Do not mark up the equation. Use the blank area of this page for your scratch work.)

1. $2x + 1 = x + 5 \leftrightarrow x =$ __________.

2. $2x = x + 18 \leftrightarrow x =$ __________.

3. $2x - 2 = x + 44 \leftrightarrow x =$ __________.

4. $3x + 1 = 0 + 22 \leftrightarrow x =$ __________.

(end of document)

(blank page – can be used as additional scratch area)

Algebra Problems 1018
(Algebra Problems: Set # 1018)

Student Name: ________________________________ Date: ____________ Score: _______

Solve for x in the following equations.
(Do not mark up the equation. Use the blank area of this page for your scratch work.)

1. $2x + 2 = x + 16 \leftrightarrow x =$ __________.

2. $2x + 3 = 0 + 25 \leftrightarrow x =$ __________.

3. $2x + 4 = x + 31 \leftrightarrow x =$ __________.

4. $2x - 5 = x + 39 \leftrightarrow x =$ __________.

(end of document)

(blank page – can be used as additional scratch area)

Algebra Problems 1019
(Algebra Problems: Set # 1019)

Student Name: ______________________________ Date: ___________ Score: _______

Solve for x in the following equations.
(Do not mark up the equation. Use the blank area of this page for your scratch work.)

1. $3x + 1 = 2x + 18 \leftrightarrow x =$ __________

2. $3x - 1 = 2x + 25 \leftrightarrow x =$ __________

3. $3x + 5 = 2x + 12 \leftrightarrow x =$ __________

4. $3x + 1 = 2x + 35 \leftrightarrow x =$ __________

(end of document)

(blank page – can be used as additional scratch area)

Algebra Problems 1020
(Algebra Problems: Set # 1020)

Student Name: ________________________________ Date: ____________ Score: _______

Solve for x in the following equations.
(Do not mark up the equation. Use the blank area of this page for your scratch work.)

1. $5x = 3x + 14 \leftrightarrow x =$ __________

2. $6x = 4x + 20 \leftrightarrow x =$ __________

3. $5x = 3x + 6 \leftrightarrow x =$ __________

4. $5x = 3x + 12 \leftrightarrow x =$ __________

(end of document)

(blank page – can be used as additional scratch area)

Algebra Problems 1021
(Algebra Problems: Set # 1021)

Student Name: ________________________________ Date: ____________ Score: _______

Solve for x in the following equations.
(Do not mark up the equation. Use the blank area of this page for your scratch work.)

1. $5x + 1 = 3x + 13 \leftrightarrow x =$ __________

2. $5x - 1 = 3x + 27 \leftrightarrow x =$ __________

3. $8x + 1 = 6x + 21 \leftrightarrow x =$ __________

4. $5x + 1 = 3x + 35 \leftrightarrow x =$ __________

(end of document)

(blank page – can be used as additional scratch area)

Algebra Problems 1022

(Algebra Problems: Set # 1022)

Student Name: ______________________________ Date: ____________ Score: _______

Solve for x in the following equations.
(Do not mark up the equation. Use the blank area of this page for your scratch work.)

1. $5x + 1 = 4x + 18 \leftrightarrow x =$ __________

2. $5x - 1 = 4x + 25 \leftrightarrow x =$ __________

3. $6x + 1 = 5x + 30 \leftrightarrow x =$ __________

4. $5x + 1 = 4x + 17 \leftrightarrow x =$ __________

(end of document)

(blank page – can be used as additional scratch area)

Algebra Problems 1023
(Algebra Problems: Set # 1023)

Student Name: ________________________________ Date: ____________ Score: _______

Solve for x in the following equations.
(Do not mark up the equation. Use the blank area of this page for your scratch work.)

1. $5x + 1 = x + 33 \leftrightarrow x =$ __________

2. $5x - 7 = x + 33 \leftrightarrow x =$ __________

3. $5x + 1 = x + 9 \leftrightarrow x =$ __________

4. $5x + 2 = x + 38 \leftrightarrow x =$ __________

(end of document)

(blank page – can be used as additional scratch area)

Algebra Problems 1024
(Algebra Problems: Set # 1024)

Student Name: ______________________________ Date: ____________ Score: _______

Example
Evaluate the following expression.
$2(x + 1)$, for $x = 7$

If $x = 7$, then $2(x + 1) = 2(7 + 1) = 2(8) = 16$

Task 1
Evaluate the following expression.
$2(x + 1)$, for $x = 3$

Task 2
Evaluate the following expression.
$2(x + 1)$, for $x = 12$

Task 3
Evaluate the following expression.
$2(x + 1)$, for $x = 8$

Task 4
Evaluate the following expression.
$2(x + 1)$, for $x = 10$

Task 5
Evaluate the following expression.
$2(x – 3)$, for $x = 10$

(end of document)

(blank page – can be used as additional scratch area)

Algebra Problems 1025
(Algebra Problems: Set # 1025)

Student Name: ______________________________ Date: ____________ Score: _______

Solve for x in the following equations.
(Do not mark up the equation. Use the blank area of this page for your scratch work.)

1. $2(x + 1) = 8 \leftrightarrow x =$ __________

2. $2(x - 1) = 14 \leftrightarrow x =$ __________

3. $2(x + 1) = 50 \leftrightarrow x =$ __________

4. $2(x + 5) = 22 \leftrightarrow x =$ __________

(end of document)

(blank page – can be used as additional scratch area)

Algebra Problems 1026
(Algebra Problems: Set # 1026)

Student Name: ______________________________ Date: ____________ Score: _______

Solve for x in the following equations.
(Do not mark up the equation. Use the blank area of this page for your scratch work.)

1. $2(x + 1) = x + 16 \leftrightarrow x =$ __________

2. $2(x - 1) = x + 38 \leftrightarrow x =$ __________

3. $2(x + 1) = x + 14 \leftrightarrow x =$ __________

4. $2(x + 1) = x + 10 \leftrightarrow x =$ __________

(end of document)

(blank page – can be used as additional scratch area)

Algebra Problems 1027
(Algebra Problems: Set # 1027)

Student Name: ______________________________ Date: ___________ Score: _______

Solve for x in the following equations.
(Do not mark up the equation. Use the blank area of this page for your scratch work.)

1. $3(x + 1) = 2(x + 3) + 7 \leftrightarrow x =$ __________

2. $3(x + 5) = 2(x + 10) + 1 \leftrightarrow x =$ __________

3. $3(x + 3) = 2(x + 6) + 6 \leftrightarrow x =$ __________

4. $3(x + 2) = 2(x + 6) + 6 \leftrightarrow x =$ __________

(end of document)

(blank page – can be used as additional scratch area)

Algebra Problems 1028
(Algebra Problems: Set # 1028)

Student Name: ______________________________ Date: ___________ Score: _______

Solve for x in the following equations.
(Do not mark up the equation. Use the blank area of this page for your scratch work.)

1. $3(x - 1) + x = 2(x + 1) + 11 \leftrightarrow x =$ __________

2. $3(x - 1) + x = 2(x + 1) + 5 \leftrightarrow x =$ __________

3. $3(x - 1) + x = 2(x + 1) + 1 \leftrightarrow x =$ __________

4. $3(x - 1) + x = 2(x + 1) + 13 \leftrightarrow x =$ __________

(end of document)

(blank page – can be used as additional scratch area)

Algebra Problems 1029
(Algebra Problems: Set # 1029)

Student Name: ____________________ Date: __________ Score: ______

Example
Evaluate the following expression.
$5x - 3(x - 1)$, for $x = 7$

If $x = 7$, then $5x - 3(x - 1) = 5(7) - 3(7 - 1) = 35 - 3(6) = 35 - 18 = 17$

Task 1
Evaluate the following expression.
$5x - 3(x - 1)$, for $x = 3$

Task 2
Evaluate the following expression.
$5x - 3(x - 1)$, for $x = 12$

Task 3
Evaluate the following expression.
$5x - 3(x - 1)$, for $x = 8$

Task 4
Evaluate the following expression.
$5x - 3(x - 1)$, for $x = 10$

Task 5
Evaluate the following expression.
$5x - 3(x - 1)$, for $x = 2$

(end of document)

(blank page – can be used as additional scratch area)

Algebra Problems 1030
(Algebra Problems: Set # 1030)

Student Name: ________________________________ Date: ____________ Score: _______

Solve for x in the following equations.
(Do not mark up the equation. Use the blank area of this page for your scratch work.)

1. $5x - 3(x - 1) = 15 \leftrightarrow x =$ __________

2. $5x - 3(x - 1) = 23 \leftrightarrow x =$ __________

3. $5x - 3(x - 1) = 19 \leftrightarrow x =$ __________

4. $5x - 3(x - 1) = 45 \leftrightarrow x =$ __________

(end of document)

(blank page – can be used as additional scratch area)

Algebra Problems 1031
(Algebra Problems: Set # 1031)

Student Name: ______________________________ Date: ____________ Score: _______

Example
Evaluate the following expression.
$x + y$, for $x = 7 \wedge y = 2$

If $x = 7$ and $y = 2$, then $x + y = 7 + 2 = 9$

Task 1
Evaluate the following expression.
$x + y$, for $x = 3 \wedge y = 12$

Task 2
Evaluate the following expression.
$x + y$, for $x = 12 \wedge y = 1$

Task 3
Evaluate the following expression.
$x + y$, for $x = 8 \wedge y = 2$

Task 4
Evaluate the following expression.
$x + y$, for $x = 10 \wedge y = 8$

Task 5
Evaluate the following expression.
$x + y$, for $x = 2 \wedge y = 18$

(end of document)

(blank page – can be used as additional scratch area)

Algebra Problems 1032
(Algebra Problems: Set # 1032)

Student Name: ________________________________ Date: ____________ Score: _______

Solve for x and y in the following equations.
(Do not mark up the equation. Use the blank area of this page for your scratch work.)

1. $y = 2x \wedge x + y = 15 \leftrightarrow x =$ __________ $\wedge\ y =$ __________

2. $y = 9x \wedge x + y = 80 \leftrightarrow x =$ __________ $\wedge\ y =$ __________

3. $y = 5x \wedge x + y = 24 \leftrightarrow x =$ __________ $\wedge\ y =$ __________

4. $y = 2x \wedge x + y = 30 \leftrightarrow x =$ __________ $\wedge\ y =$ __________

(end of document)

(blank page – can be used as additional scratch area)

Algebra Problems 1033
(Algebra Problems: Set # 1033)

Student Name: ______________________________ Date: ____________ Score: _______

Solve for x and y in the following equations.
(Do not mark up the equation. Use the blank area of this page for your scratch work.)

1. $y = 2x \wedge y - x = 8 \leftrightarrow x =$ __________ $\wedge y =$ __________

2. $y = 9x \wedge y - x = 16 \leftrightarrow x =$ __________ $\wedge y =$ __________

3. $y = 5x \wedge y - x = 36 \leftrightarrow x =$ __________ $\wedge y =$ __________

4. $y = 2x \wedge y - x = 13 \leftrightarrow x =$ __________ $\wedge y =$ __________

(end of document)

(blank page – can be used as additional scratch area)

Algebra Problems 1034
(Algebra Problems: Set # 1034)

Student Name: ______________________________ Date: ____________ Score: _______

Solve for x and y in the following equations.
(Do not mark up the equation. Use the blank area of this page for your scratch work.)

1. $x + y = 8 \wedge x - y = 2 \leftrightarrow x =$ __________ $\wedge\ y =$ __________

2. $x + y = 12 \wedge x - y = 8 \leftrightarrow x =$ __________ $\wedge\ y =$ __________

3. $x + y = 13 \wedge x - y = 11 \leftrightarrow x =$ __________ $\wedge\ y =$ __________

4. $x + y = 5 \wedge x - y = 3 \leftrightarrow x =$ __________ $\wedge\ y =$ __________

(end of document)

(blank page – can be used as additional scratch area)

Algebra Problems 1035
(Algebra Problems: Set # 1035)

Student Name: ________________________________ Date: ____________ Score: _______

Solve for x and y in the following equations.
(Do not mark up the equation. Use the blank area of this page for your scratch work.)

1. $3x + 2y = 30 \wedge 2x + 3y = 35 \leftrightarrow x =$ __________ $\wedge\ y =$ __________

2. $2x + 5y = 36 \wedge 3x - 2y = 16 \leftrightarrow x =$ __________ $\wedge\ y =$ __________

3. $4x + 2y = 14 \wedge 2x + 9y = 47 \leftrightarrow x =$ __________ $\wedge\ y =$ __________

4. $2x + 2y = 22 \wedge 3x - 4y = 19 \leftrightarrow x =$ __________ $\wedge\ y =$ __________

(end of document)

(blank page – can be used as additional scratch area)

Table: Prime Factorizations up to 100

	$21 = 3 \times 7$	$41 = 41$	$61 = 61$	$81 = 3^4$
$2 = 2$	$22 = 2 \times 11$	$42 = 2 \times 3 \times 7$	$62 = 2 \times 31$	$82 = 2 \times 41$
$3 = 3$	$23 = 23$	$43 = 43$	$63 = 3^2 \times 7$	$83 = 83$
$4 = 2^2$	$24 = 2^3 \times 3$	$44 = 2^2 \times 11$	$64 = 2^6$	$84 = 2^2 \times 3 \times 7$
$5 = 5$	$25 = 5^2$	$45 = 3^2 \times 5$	$65 = 5 \times 13$	$85 = 5 \times 17$
$6 = 2 \times 3$	$26 = 2 \times 13$	$46 = 2 \times 23$	$66 = 2 \times 3 \times 11$	$86 = 2 \times 43$
$7 = 7$	$27 = 3^3$	$47 = 47$	$67 = 67$	$87 = 3 \times 29$
$8 = 2^3$	$28 = 2^2 \times 7$	$48 = 2^4 \times 3$	$68 = 2^2 \times 17$	$88 = 2^3 \times 11$
$9 = 3^2$	$29 = 29$	$49 = 7^2$	$69 = 3 \times 23$	$89 = 89$
$10 = 2 \times 5$	$30 = 2 \times 3 \times 5$	$50 = 2 \times 5^2$	$70 = 2 \times 5 \times 7$	$90 = 2 \times 3^2 \times 5$
$11 = 11$	$31 = 31$	$51 = 3 \times 17$	$71 = 71$	$91 = 7 \times 13$
$12 = 2^2 \times 3$	$32 = 2^5$	$52 = 2^2 \times 13$	$72 = 2^3 \times 3^2$	$92 = 2^2 \times 23$
$13 = 13$	$33 = 3 \times 11$	$53 = 53$	$73 = 73$	$93 = 3 \times 31$
$14 = 2 \times 7$	$34 = 2 \times 17$	$54 = 2 \times 3^3$	$74 = 2 \times 37$	$94 = 2 \times 47$
$15 = 3 \times 5$	$35 = 5 \times 7$	$55 = 5 \times 11$	$75 = 3 \times 5^2$	$95 = 5 \times 19$
$16 = 2^4$	$36 = 2^2 \times 3^2$	$56 = 2^3 \times 7$	$76 = 2^2 \times 19$	$96 = 2^5 \times 3$
$17 = 17$	$37 = 37$	$57 = 3 \times 19$	$77 = 7 \times 11$	$97 = 97$
$18 = 2 \times 3^2$	$38 = 2 \times 19$	$58 = 2 \times 29$	$78 = 2 \times 3 \times 13$	$98 = 2 \times 7^2$
$19 = 19$	$39 = 3 \times 13$	$59 = 59$	$79 = 79$	$99 = 3^2 \times 11$
$20 = 2^2 \times 5$	$40 = 2^3 \times 5$	$60 = 2^2 \times 3 \times 5$	$80 = 2^4 \times 5$	$100 = 2^2 \times 5^2$

(end of document)

(blank page)

Blank Prime Factorizations Table

Student Name: ________________________________ Date: ____________ Score: _______

Fill in the prime factorizations for the given numbers. Write the prime factorizations in standard form (smallest primes first, and using exponents for repeated prime factors).

	21 =	41 =	61 =	81 =
2 =	22 =	42 =	62 =	82 =
3 =	23 =	43 =	63 =	83 =
4 =	24 =	44 =	64 =	84 =
5 =	25 =	45 =	65 =	85 =
6 =	26 =	46 =	66 =	86 =
7 =	27 =	47 =	67 =	87 =
8 =	28 =	48 =	68 =	88 =
9 =	29 =	49 =	69 =	89 =
10 =	30 =	50 =	70 =	90 =
11 =	31 =	51 =	71 =	91 =
12 =	32 =	52 =	72 =	92 =
13 =	33 =	53 =	73 =	93 =
14 =	34 =	54 =	74 =	94 =
15 =	35 =	55 =	75 =	95 =
16 =	36 =	56 =	76 =	96 =
17 =	37 =	57 =	77 =	97 =
18 =	38 =	58 =	78 =	98 =
19 =	39 =	59 =	79 =	99 =
20 =	40 =	60 =	80 =	100 =

(end of document)

(blank page – can be used as additional scratch area)

Algebra Problems 1036
(Algebra Problems: Set # 1036)

Student Name: ______________________________ Date: ____________ Score: _______

Solve for x in the following equations.

1. x is the largest prime less than 15 $\leftrightarrow$ x = __________.

2. x is the largest prime less than 25 $\leftrightarrow$ x = __________.

3. x is the smallest prime greater than 14 $\leftrightarrow$ x = __________.

4. x is the smallest prime greater than 30 $\leftrightarrow$ x = __________.

5. x is the largest prime less than 10 $\leftrightarrow$ x = __________.

6. x is the smallest prime greater than 50 $\leftrightarrow$ x = __________.

7. x is the smallest prime greater than the largest prime less than 20 $\leftrightarrow$ x = __________.

8. x is the largest prime less than the smallest prime greater than 20 $\leftrightarrow$ x = __________.

(end of document)

(blank page – can be used as additional scratch area)

Made in the USA
Middletown, DE
13 May 2018